Genre > Nonfiction

W9-AWH-941

? Essential Question
What are characteristics of a desert biome?

Building a Biome

by Anna Prokos

Home in the Biome

The sunshine heats up the desert. A tortoise crawls under a boulder. A Colorado River toad jumps into a puddle. In the wild, these **species** live in the Sonoran Desert in Arizona.

There is another place you can find these animals. It is in the Indianapolis Zoo's Deserts Biome. Here, the zoo has made an indoor desert to show the living things that live in desert **biomes**.

The zoo's Deserts Biome has features of three real deserts around the world.

The Sonoran Desert covers parts of California, Arizona, and Mexico.

Chapter 1

Deserts in the United States

The United States has four major deserts. They are in the southwestern part of the country.

Deserts around the world have plants and animals that are endemic to that desert. That means they can only survive in that specific desert.

Other species are native to the desert. They are found mostly in that desert habitat. However, they can also survive in other desert environments as well.

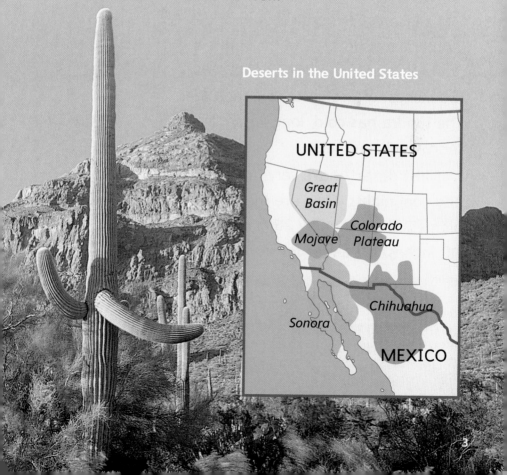

Deserts in the United States

UNITED STATES

Great Basin

Colorado Plateau

Mojave

Sonora

Chihuahua

MEXICO

Biomes Rule

A biome is an environment where plants and animals live. There are six kinds of biomes on Earth. They are tropical rain forest, grassland, deciduous forest, taiga, tundra, and desert.

Rain forest

Each biome has a different climate. Tropical rain forests are hot all year. Rain forests can have up to 457 centimeters (180 inches) of rain in a year. Grasslands have cool winters and hot summers. Deciduous forests have four seasons. The taiga is very cold. Winters are snowy, and the summers are cool. It gets about 51 centimeters (20 inches) of **precipitation** each year. The tundra has cold, long winters and short, cool summers.

Grassland

Deciduous Forest

Most deserts have very hot days and cool nights. They are the driest biomes. They get only 25 centimeters (10 inches) of rain or fewer in a year.

Taiga

Tundra

Desert

Chapter 2

Desert in a Zoo

The state of Indiana is not close to any deserts. Even so, you can find desert plants and animals living in the Deserts Biome in the Indianapolis Zoo.

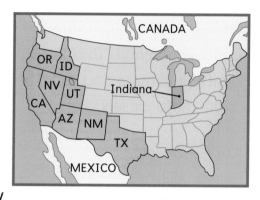

A lot of people like to visit the Deserts Biome. They feel like they are in a real desert. That is because the staff at the zoo has worked hard to make the Deserts Biome like the natural desert environment.

Desert lizards are at home in the Deserts Biome.

Building a Biome

When the zoo decided to make a desert biome, it sent some of its staff to a real desert. Scientists who work at the zoo spent two weeks in the Sonoran Desert in Tucson, Arizona. They trekked through the desert to study the plants and animals that lived there. They took notes about the climate and soil. The scientists observed their surroundings closely.

The Sonoran Desert

What Is in a Biome?

The plants and animals in a biome are suited, or adapted, to living in their environment.

Plants are **producers**. They make their own food. Animals are **consumers**. Some animals eat the plants. Some animals eat the animals that eat plants.

Producer

When plants and animals die, **decomposers** break down the organisms. They become material that is rich in nutrients. Plants need this material to grow.

Consumer

Scientists working for the zoo paid attention to the plants and animals in the desert. Then they studied the desert's **food chains**.

Decomposer

Desert Food Chains

Every biome has several food chains. They all begin with sunlight. Plants use the Sun's energy to make food. Then animals eat the plants. The energy is passed to them. The same thing happens when an animal eats another animal that eats plants. Next, the energy passes to decomposers.

There are many food chains in a biome. Some food chains have the same plants and animals. Food chains that are linked together are part of a food web.

Deserts are dry, but there are plants and animals that can live there. The plants and animals that live in a specific desert environment need each other to survive.

Energy passes from producers (plants) to consumers (animals) to decomposers (bacteria and fungi) in a food web.

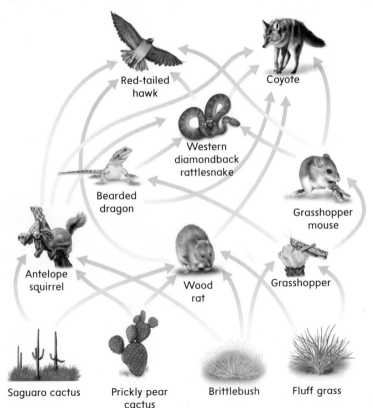

Red-tailed hawk

Coyote

Western diamondback rattlesnake

Bearded dragon

Grasshopper mouse

Antelope squirrel

Wood rat

Grasshopper

Saguaro cactus

Prickly pear cactus

Brittlebush

Fluff grass

Desert Adaptations

Desert plants and animals have **adaptations** that allow them to survive. Adaptations are characteristics that help an animal survive in its environment.

The Colorado River toad and desert tortoise spend most of the day in cool underground burrows. Other animals are active during the day and stay in burrows at night.

It does not rain a lot in the desert, so plants need to save water. They store it in their stems.

Nights in the desert are cool. Many animals are active at night instead of during the day.

This desert fox has large ears that help it stay cool.

Rocks and Other Features

The staff at the zoo spent a lot of time studying rocks in the Sonoran Desert. They made rubber molds of the rocks they found. Then they made models of the rocks to put in the exhibit.

The floor of the Deserts Biome exhibit is made of gravel. The rocks were painted to match the gravel. They were placed around the exhibit to make visitors feel like they were in a real desert.

The rocks form a canyon and river. In the Sonoran Desert, animals live in the Colorado River. They rest in the shade of canyons.

The Sonoran Desert landscape

Chapter 3
Home in the Dome

In the Deserts Biome, small lizards run and birds fly around. The exhibit is enclosed in a geodesic dome. The dome keeps animals from escaping.

A geodesic dome is a building with a pattern of triangles. The triangles give it a strong structure. The dome in the Indianapolis Zoo is made out of see-through material. It lets light enter the Deserts Biome. It also protects the plants and animals from temperature changes, rain, and pollution.

A geodesic dome formed of triangles

Weather in the Dome

The plants and animals in the Deserts Biome need desertlike temperatures to survive. The zoo uses heating and air-conditioning to keep the right temperatures. The temperature is made a little cooler in the winter.

The animals need the lowered temperature to hibernate just as they would in the wild.

On rainy or cloudy days, some rocks and mats in the biome are heated to a high temperature. This is like the warmth of sunlight. The animals feel like they are in their natural environment.

Many coyotes call the desert home.

Horned lizard

Plans for Plants

Plants play a major part in the Deserts Biome, too. The workers at the zoo spent a lot of time studying desert plants. They found information to identify out which plants would survive well in a desertlike zoo environment.

The Deserts Biome has desert roses, baobab trees, ponytail palms, boojum trees, and other plants. The workers tried to include as many desert plants as possible. It was hard to find desert plants that would survive in Indianapolis. They knew these plants would survive.

Cactus flowers bloom in deserts.

Some plants adapt well to desert life.

Keeping Cool

Some animals need very low temperatures to hibernate. The desert tortoise is one of these animals. Zookeepers place the tortoises in a special refrigerator away from the exhibit so that they can hibernate. The refrigerator keeps the turtles' environment at a cool 13 degrees Celsius (55 degrees Fahrenheit) all winter. The tortoises use very little oxygen during hibernation.

Zookeepers place desert tortoises like this one in a special refrigerator so they can hibernate.

Secret Hiding Places

Lizards and other desert animals are not used to a lot of visitors in their habitat. That is why the zoo worked hard to help the animals feel safe. There are 75 tubes hidden in the zone. The tubes are hidden inside large rocks in the exhibit. These special tubes give shelter and protection to lizards and other small reptiles. Special tubes are not part of a natural desert, but they are important in the Deserts Biome.

Gila monsters live in the Deserts Biome.

Turn Up the Heat

The African desert known as the Sahara is about three times the size of Alaska! The Sahara is one of the hottest deserts on Earth. Temperatures can reach 58 degrees Celsius (136 degrees Fahrenheit)! In the United States, the Sonoran Desert is the hottest. It reaches about 49 degrees Celsius (120 degrees Fahrenheit). Yet not all deserts are hot.

The Sahara is the second-largest desert in the world.

A desert does not need to be hot. Scientists describe a desert as an environment that receives fewer than 25 centimeters (10 inches) of rainfall each year. That means that the entire continent of Antarctica is a cold desert! Antarctica is covered with snow and ice, but it is a very dry place. Any precipitation that falls is in the form of snow. If you took a measurement of the snow in Antarctica, it would equal about 5 centimeters (2 inches) of rainfall each year. That is not a lot of water!

The largest desert in the world is Antarctica.

Plants of the Deserts Biome

Choosing the right plants for the Deserts Biome was hard work. These plants make the Deserts Biome look unique, or special.

Desert rose: This flowering plant can live for hundreds of years in an African desert.

Ponytail palm: Native to arid regions of Mexico, the trunk of this plant is shaped like an onion.

Cardon: One of the largest of all cactuses, this cactus can weigh 22.5 metric tons (25 tons)!

Baobab tree: You can find this tree in the savannas of Africa and India, mostly around the equator.

Kokerboom: This plant grows in the arid regions of South Africa. It attracts birds, insects, and baboons. These animals eat the tree's bright yellow flowers.

Euphorbia ingens: This tree grows in areas of Africa. It has a poisonous sap. In the wild, people start a fire around this tree before cutting it down. The fire keeps the sap from spreading.

Alluaudia ascendens: This plant has spikes as well as small leaves that hold water. It is from the African country of Madagascar.

Boojum tree: When it rains, rounded leaves quickly grow from the thin branches along the stem.

Strange Looks

Boojum trees are native to the Sonoran Desert. The boojum tree has an unusual shape. It looks like a giant upside-down carrot. These trees can live up to 300 years!

Boojum trees are made of wood that is not very strong. That is why they break in many places. The trunks branch, or split near the top. They sometimes have strange shapes.

White flowers grow at the tips of the boojum's trunk. Boojum trees have a thick layer under their outer skin. This layer keeps them alive for years, even with little rainfall. The humidity in the air allows the boojum tree to flower and spread seeds. The boojum tree grows well in the controlled environment of the Deserts Biome.

Chapter 4

Sonoran Desert vs. Deserts Biome

Bearded dragon

Zoos give visitors a chance to observe how plants and animals survive naturally in the world. Zoo exhibits are often different than natural habitats. The staff at the zoo tries to use as many natural details as possible. What are the similarities and differences of the Deserts Biome? Visitors can compare the two environments to see what it takes to keep a zoo desert exhibit similar to the real thing.

The Desert Dome at the Omaha zoo

Sonoran Desert	Deserts Biome
Location: Southern Arizona, stretching into southeast California, Baja California, and Sonora, Mexico	**Location:** Indianapolis Zoo, Indianapolis, Indiana
Number of visitors a year: about 550,000	**Number of visitors a year:** 1,000,000
• Gets cooler when the Sun goes down. Night temperatures may even be near freezing!	• Gets cooler when the Sun goes down. Zoo staff turns up air conditioners to keep the biome cool at night.
• Plants and animals are part of a food web.	• Plants and animals are not part of a natural food web. Zookeepers feed animals each day.
• Predators, such as rattlesnakes, hunt for food at night.	• Animals do not need to hunt for food.
• People who trek through the desert can get sunburned.	• The dome's glass let in the warmth of the Sun and filters out harmful ultraviolet rays.

Chapter 5

Species of the Deserts Biome

All the animals in the Deserts Biome live naturally in the Sonoran Desert, in African deserts, and in dry areas of Australia.

Giant plated lizard

Reptiles

Desert iguanas: These iguanas climb into bushes to escape heat but are active in temperatures as hot as 46 degrees Celsius (115 degrees Fahrenheit).

Rhinoceros iguanas: They lay in the Sun to become active but return to their burrows for a cool rest.

Giant plated lizards: These lizards live in Africa. Sometimes, they eat smaller lizards or baby tortoises.

Grand Cayman blue iguanas: Only a small number of these animals are left in the wild in Grand Cayman.

Bearded dragons: Native to arid areas of Australia, they can regulate, or control, their body temperature through evaporation.

Desert tortoises: These tortoises spend 95 percent of their life in burrows.

Pancake tortoises: Their flat shells let them crawl under rocks for protection from the heat.

Pancake tortoise

Amphibians

Colorado River toads: These toads make poison to protect them from hungry predators.

Colorado River toad

Birds

Diamond doves: Native to Australia, they are one of the smallest doves.

Cordon bleu finches: The beak of these birds is adapted to peck at the grasses and trees of the desert.

Saudi Arabian sand partridges: They live in the semi-desert areas of Saudi Arabia.

Scaled quails: They live in the Sonoran Desert and other areas of the United States and Mexico.

Orange-cheeked waxbills: These African birds eat seeds on tall grasses.

Red fodys: These bright red birds live in Madagascar.

Gambel's quails: These birds are 28 centimeters (11 inches) long.

Scaled quail

Hungry Animals

In their natural biomes, animals hunt for food or eat plants in their environment. In a zoo, animals are fed a balanced diet. This is how much food all the animals at the Indianapolis Zoo need each day.

| 91 kilograms (200 pounds) fruits and vegetables | 181 kilograms (400 pounds) meat | 272 kilograms (600 pounds) grain | 100 bales hay and bedding |

Saving Species

The Indianapolis Zoo leads efforts to learn more about endangered species, both in the wild and at the zoo. Two researchers study lemurs in the wild in Madagascar. Then they compare that information to data gathered from the ring-tailed lemurs at the zoo. They hope to learn more about how humans impact wild populations. The zoo also has a strong breeding program involving many kinds of animals. In 2014 alone, an amur tiger, three Caribbean flamingos, four warthogs, and a ring-tailed lemur were born.

The zoo is recognized around the world as a research center and a place for breeding rare desert reptiles. One of the zoo's breeding programs focuses on Grand Cayman blue iguanas. They have almost died out in their natural habitat.

It is difficult to hatch Grand Cayman blue iguana eggs in a zoo, but the Indianapolis Zoo has been successful two times. In 2002, seven blue iguanas hatched at the zoo. In 2005, two Grand Cayman blue iguanas hatched at the zoo. Other zoos in the United States are also breeding blue iguanas.

The Blue Iguana Recovery Programme of Grand Cayman works to protect the blue iguanas in their natural habitat. A zoologist from the Indianapolis Zoo helped to build special fences to keep the blue iguanas safe in Grand Cayman.

These programs are working to save the Grand Cayman blue igauana. In 2002, there were only 10 to 25 blue iguanas in the wild. In 2013, there were 750.

A breeding program at the zoo is helping to save the Grand Cayman blue iguana.

Beyond the Biomes

When the Indianapolis Zoo built the Deserts Biome, it had **conservation** in mind. The goal of the Deserts Biome, like the zoo's other exhibits, is to educate visitors about plant and animals in a real environment. The biomes in the zoo are as close as you can get to actual environments without traveling to them!

In the future, the Indianapolis Zoo plans to add more ways for people to experience the wild natural world of plants and animals.

The Indianapolis Zoo empowers people and communities, both locally and globally, to advance animal conservation.
— Indianapolis Zoo Mission Statement

Summarize

Use important details from the text to summarize *Building a Biome.* Your graphic organizer may help you.

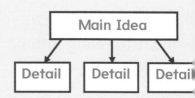

Text Evidence

1. How do you know that *Building a Biome* is a nonfiction text? Identify the text features that tell you this.

2. Reread this book with a group. Take notes on each chapter. Take turns identifying key details about ways in which the Indianapolis Zoo has worked to make the Deserts Biome as real as possible. MAIN IDEA AND KEY DETAILS

3. What is the meaning of the word *live* on page 24? What is another meaning for the word *live?* What clues in the text show you which meaning to use on page 24? HOMOGRAPHS

4. Use different resources to find out about a different biome. You might want to find out about a tundra, taiga, or grassland. Write a report about the biome you select. What are its characteristics? What types of plants and animals can be found there? Be sure to include a main idea in your report and support it with key details. WRITE ABOUT READING

Compare Texts

Find out how Wyatt uses an app to find the best biome for different animals.

Wyatt, King of the Biomes

"Wyatt, is that phone permanently attached to your hand?" "No, Mom." Wyatt called back, never moving his eyes from the screen, "It's a new simulation that requires the identification of an animal's biome based on its characteristics. I'm on Level 9! If I can distinguish one more animal, I'll be King of the Biomes! Can dinner be delayed ten minutes?"

Wyatt examined the bright screen. It displayed a specific set of characteristics. "Eight feet tall and darkly colored. Its ears are large and round. but its face is narrow." Wyatt scratched the side of his head.

Additional characteristics flowed across the screen. "This animal possesses ivory that is somewhat pink and has tusks that are not curved. This allows the animal to move more easily about its biome since straight tusks help it avoid getting caught in vines and underbrush." Wyatt's eyes popped wide open. Not many animals had tusks! Wyatt's thought to himself, "I should be able to narrow this down!

Wyatt continued examining the information. "This animal maintains a small family group and ranges a territory of a thousand square miles. It is shorter and darker in color than its much bigger savanna-living cousins. Its smaller size allows it to move around in their dense biome. It is important for the germination of seeds of plants for which its biome was named." A bright smile crossed Wyatt's face. He would puzzle it out during dinner.

"What? No dessert?" "No, Mom! I've had an interesting thought," said Wyatt. He carefully studied the characteristics again, thinking "Underbrush, vines, smaller frames that allows better movement. Straight, pinkish tusks, is a member of a smaller family group but its cousins live in the savanna!" Undergrowth and vines are not usually features of the savanna! It's smaller than its cousins at eight feet! It moves in families. An elephant that has straight pinkish tusks. The smallest of the species? The African Forest Elephant! Wyatt used his thumbs to type Biome: Tropical Rainforest. The flashing screen displayed, "Wyatt, King of the Biomes!"

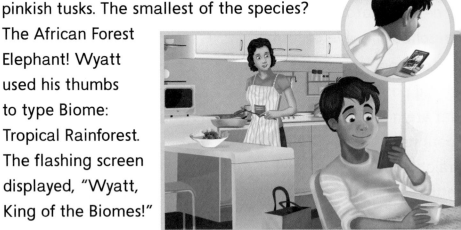

Make Connections

How can you identify the best biome for an animal to live in? TEXT TO TEXT

Glossary

adaptations *(ad-dap-TAY-shuhn)* features that help the survival of a species *(page 9)*

biomes *(BIGH-ohms)* Earth's ecosystems, including climate, soil, plants, and animals. *(page 2)*

conservation *(kon-suhr-VAY-shuhn)* efforts to protect natural resources *(page 27)*

consumers *(kuhn-SEW-muhrs)* animals that eat plants or other animals *(page 7)*

decomposers *(dee-kuhm-POH-zuhr)* organisms that break down dead plants and animals into materials that enrich soil *(page 7)*

food chain *(FEWD CHAYN)* a model of how energy in food is passed among organisms in an ecosystem *(page 7)*

precipitation *(prih-sip-uh-TAY-shuhn)* falling water in the form of rain, sleet, hail, or snow *(page 4)*

producers *(proh-DEW-surs)* green plants or one-celled organisms that can make their own food *(page 7)*

species *(SPEE-sheez)* a group of organisms that can reproduce its own kind *(page 2)*

Index